'Black Celestial Bodies' (2016)
Printed by CreateSpace, An Amazon.com Company

---

**-US COPYRIGHT OFFICE, eCO Filing 2016**

1-3312696201 Black Celestial Bodies' *Literary Work.*

Related work:
1-3248930091 *Performing Arts.* Music composition for Black Celestial Bodies' stage and/or screenplay. Author and piano interpreter: Sergio Pablo Sánchez Cordero Dávila.

---

**-INDAUTOR MEXICO (Mexican Copyright Office)**

a) Black Celestial Bodies, Literary work
Reg. 03-2016-042710101900-01

Related Work:
b) *Performing Arts.*Instrumental music for the stage and/or screenplay representation. Composer and pianist by the same author.
Reg. 03-2016-042710091200-01

---

**About the Self-Publishing Author**
New York University Master of Science, class of 1978.

Basic science publications in microbiology and clinical periodontology. Holds patents in the USA and Mexico.

New York University College of Dentistry, Postgraduate in Periodontology 1975-1977.

Other published books and additional info on page 65.

**Covers:**

**Star-Forming Region LH 95 in the Large Magellanic Cloud**

To extend your knowledge about LH 95 please go to the following link:
**http://hubblesite.org/gallery/album/star/pr2006055a/**

In deep gratitude with the **Hubble Heritage Team** for allowing me to reproduce this astonishing interstellar photograph. Besides its extraordinary beauty, this intense star-forming region stands not only as the covers of my book, but as a metaphor that beholds the implicate concept of a Stories' Universe described in the literary work ahead.

It took 160,000 light years for the image to reach the earth; its constituent photons stayed virtually unchanged during their long journey until they finally went through our eyes; the retinal cells welcoming the physical cosmos to another universe, the human brain. It is here where our consciousness dresses it with poetic thoughts and unravels its mysteries under the lens of science.

The starred heaven now holds a place I can call home: **LH 95**

The Author

# "Black Celestial Bodies"

## The Story of Stories

## A Metaphor of Human Consciousness

by

Sergio Sánchez Cordero

Pablo, the Scientist
Santiago, the Artist
My Self
Inhabiting the same story
Confluence of minds creating realities

...the cosmic light is now moving through another universe, the human mind, where the sensed world and imagined reality ink-skate together like figure skaters on the white page, dancing to the rhythm, melody and harmony of language...

## FOREWORD

In 1905 Einstein published four breakthrough papers: on the photoelectric effect (Nobel Prize in 1927), Brownian motion, special relativity, and the equivalence of mass and energy (E=mc2). It was called his 'miracle year'. Along with a host of luminaries in theoretical physics like Planck, Bohr, De Broglie, Heisenberg, Schrödinger and others, the way for nuclear power was paved, time was no longer an absolute in the universe, a space-time continuum was an essential part of the universe's fabric that helped explain gravity, quantum mechanics evolved. *Our concept of reality changed radically;* mathematics was the tip of the iceberg representing the universe's wholeness and human life would never be the same.

While the discoveries in physics captured front-page news, other fields of knowledge were just beginning. In 1906 the seeds of neuroscience were first planted. Ramón y Cajal and Camillo Golgi shared the Nobel Prize in Physiology or Medicine for their work on the structure of the nervous system. Ramón y Cajal was the conceptualizer of what became the *neuron doctrine*, which holds that the nervous system is made up of individual cells which connect to each other; named neuron, these cells have a central body or soma, a dendritic tree (input) an axon (output) and connect to other neurons by means of astonishing minuscule structures called synapses. It is precisely on the versatility of these synapses that the plasticity of the brain rests upon.

Then came one of the most fascinating discoveries in medical science. Having as a laboratory model the giant axon of the squid, Hodgkin and Huxley not only unraveled the output functioning of the neuron, but mathematically arrived at the equation that *described and predicted* in detail this specific function (Nobel Prize 1963). For the first time the action potential or 'spike' of a neuron was graphically recorded. The nervous system's bioelectrical unit had been revealed.

Now we know that the human brain is constituted by around *86 billion of these bioelectrical cells* and their connections reach trillion numbers. Neurons display a great variety of shapes, sizes, and specialize in specific functions. Their complete classification stands as a formidable task, while the Human Connectome Project focuses on connections and neuronal circuits, unraveling their complexity and interdependence.

The HUGO (Human Genome Project), continues to be the role model of the required international coordinated efforts.
The US Brain Initiative and the European Brain Project, among others, talk about government and society commitment in this regard.
Cognitive neuroscience, cognitive psychology, neurophysics, neurophilosophy, computational neuroscience, etc., are only a few examples of related evolving fields.

Cutting-edge technologies like Optogenetics and Magnetic Resonance Imaging are some of today's available powerful neuroscience tools, giving us an astonishing insight on the intricate structure of the brain and how it functions.

Just like it happened in the early 1900s with the physics revolution, neuroscience will change *our concept of reality in a radical manner*, but this time the change also includes the notion of *my self, your self and our selves*. It all revolves around a central question, what is consciousness?

Articulated language is a *foremost expression* of consciousness. Images, signs, symbols and sounds, represent not only a sensed world but also an imagined reality. *It is here where Black Celestial Bodies begins*. The present approach to consciousness is *surrealistic and metaphorical*. The elusive and ethereal existence of the mind's *Narrating Voice* is described as a *surreal wave,* resembling the structure and properties of an *electromagnetic wave* (e.g. visible light). The electrical activity of the brain can indeed be recorded as a series of wave-tracings, which actually constitute the EEG (ElectroEncephaloGram) that has specific medical applications. One key pioneer in this regard was Hans Berger, a German physiologist and psychiatrist, who recorded the first human brain EEG back in 1924.

As the story unfolds, the *quantum double-slit experiment* takes place, which is crucial in demonstrating the dual particle-wave nature of light. It will also help understand the *Narrating Voice's surreal wave*. We advise the reader to search for YouTube-videos, where this amazing experiment is explained at all levels of knowledge, for it can be life-changing.

The *fantastic allegory* of envisioning the Final Period as the Black Hole of an imaginary Stories' Universe, placed after the last word on the white page, is the fabric that holds the whole story together. Let SAM himself guide you through this cosmos...

The Author
Mexico City 2016

***

# The Story of Stories

The Big-Bang signaled the creation of the universe, of matter and of electromagnetic radiation that includes visible light; simultaneously and entangled with this physical universe, a Narrating Voice created a parallel cosmos, where human consciousness finds an existence through articulated language in the stories' universe. Every moment of every story real or imagined, is being inked down on a white page as it happens creating an ocean of signs, symbols, drawings and words. Bound by a time frame, the river of all physical beings finally arrives to this ocean. It joins the wholeness of human consciousness constituted by billions of thoughts that have found refuge here, its dynamic waves carrying their Narrating Voices to every shore.

Somewhere in that cosmos stands a Laboratory of Science and Alchemy, a fascinating place packed with a diversity of objects, models, images and devices; beakers all sizes and shapes of ongoing experiments, human brain models and brain microscopy-slides showing neurons with dendritic and axon trees in all imaginable design and layouts, blackboards full of chemical reactions, wave-function equations, general and special relativity equations amidst poems and philosophical reflections.

SAM, the Science-Alchemist-Magician founder of the laboratory, is performing laboratory bench work and while a distillation essay is running, he takes a chalk in his hand, stands in front of the blackboard at a distance...looks at the equations for a minute or two...modifying some of them here and there and making notations. Then he goes to the brain model and holds it in his hand. He slowly turns his head around and looks outside through the window at the cosmos and thinks for himself...

—This is what the mind of human beings is made of: an interwoven tapestry of living matter and consciousness. Tangible and intangible co-existing in the same space-time continuum. How is this possible? The fascinating view of a starred heaven at night holds the answers. The light of those stars has been traveling for millions of years across the universe unchanged…until it goes through our observing eyes and hits the retina.

—It is then, and only then, that a unique metamorphosis occurs made only possible by the plasticity of billions of neuronal synaptic connections in the brain: those very photons of light trigger a cascade of biochemical reactions leading to spiking neurons and a chain of events that allows the physical world to become images. The other senses follow in the orchestration of a symphony of neuron circuitry involving the associative cortex, and consciousness awakes.
A brain waiting for the cosmic light…to rise as a unique being in the whole physical universe.

—The true magic, however, is more elusive and far more transcendental; a Voice within our minds arises and becomes the narrator not only of what we experience, but of ourselves. A piano soloist playing masterfully and directing the whole orchestra at the same time; an entity that is immaterial and ethereal in nature, perhaps contained in a mathematical expression of neuronal-network interactions that reaches far beyond the electrochemical properties of neurons and their synaptic connections.
Being inaudible to others and with an aural perception by consciousness, the Voice is kept as the greatest personal secret in everybody's mind.

—Like a butterfly coming out of its cocoon and unfolding its wings, ready to become airborne, consciousness leaves its physical existence behind. The Voice is now in a quantum state, intangible, cannot be circumscribed and it takes the nature of light itself. As light, it has its origin in matter, but after being emitted both become massless, *they are now in another plane above matter capable of generating extraordinary realities.* Light made possible life on earth and the Narrating Voice developed an articulated language that constitutes the highlight of the human brain; it is precisely here where spontaneous new visions of reality take place, giving way to science, art and literature.

In sharp contrast as it happens with their sources, light and Narrating Voice cannot be contained within a time-frame. Long after its original sun is gone, light will keep-on crossing the universe eternally; the Narrating Voice in turn, will inhabit the stories' universe for ever, long after its originating brain has ceased to exist.

—In more than a sense, the cosmic light has metamorphosed into another form of light, the Narrating Voice, preserving its intrinsic structure*. The Voice embodies two intertwined 'fields' simultaneously traveling together as a radiating wave, just as electro-magnetic waves do; but in the case of the Voice one of its fields carries the reality of the physical world as perceived by the senses, and the other of imagined reality; they ink-skate together like figure skaters on the white page, dancing to the rhythm, melody and harmony of language.

—White page and consciousness share a co-dependent existence that reflects human nature. The Narrating Voice is not only the driving force behind the ink that writes a story or elaborates a painting; its ethereal nature surfaces again as it becomes independent from the ink itself.

Words, sounds and symbols represent not only the physical world, but also an imagined reality. Part of this imagined reality are the white pages themselves upon which language will find an existence in this stories' universe cosmos; they share a surreal state of existence with other entities essential to man, like the elusive number 0 which gave way to the decimal system; a fundamental pillar of civilization. White pages and the number 0, *are imaginary and real at the same time.*

---

* To better understand the *surreal wave nature* of the Narrating Voice, we advise the reader to search for *'the nature of light'* in YouTube-videos, where it is explained at all levels of knowledge.

—In the cosmos of the physical universe, a sun/star will go through a life-cycle: it forms, develops and after a determined period of time, when it runs out of nuclear fuel, its internal pressure is insufficient to resist the sun's own gravity, resulting in a 'gravitational collapse' or implosion towards its center constituting a Black Hole. Its enormous density will devour everything around it including light, space and time. Part of its mass however will radiate as a gravitational wave spreading through the vastness of the space surrounding it, letting every corner of the universe know that the end of a star has occurred.

—In the parallel cosmos of the stories' universe, a unique type of solar system evolved: the *story-solar-system*. Its origin starts with an ethereal cosmic dust of human consciousness forming the sun that will light that particular story. The story inhabits and unfolds within an orbiting planet.

—As it happens in the physical cosmos, the collapse of its sun, signals the end of the story. A whole book compressed to just a tiny dot placed after the last letter of the last word. This is the Final Period and constitutes the Black Hole of the stories' universe. Of an extraordinary density, its enormous gravity-pull devours everything in its neighborhood; the orbiting planet where the story developed with its corresponding light, space and time. The binding forces that can keep it in such a dense state are the mighty Forces of Fate. Each story was determined to start, unfold and end in a precise way. The Final Period stands as a Black Titan guarding the story and its destiny for eternity. Only the emitted Narrating Voice escapes the Final Period; it is of a timeless nature and keeps forming part of the wholeness of the stories' universe cosmos forever.

—Every single moment of every single story, lived or imagined, is being written by the Narrating Voice on the white page exactly as it happens in real time: two entangled universes mirroring each other. Finally, it is precisely on the white page where the story will find its *only* existence, starting with the first word and ending with a Final

Period. All physical existence is bound to become written words. The  stories' universe is dotted with billions of these *story-solar-systems*, organized in galactic clusters.

The end of a single story is an event that does not go unnoticed by the rest of this universe. As the Final Period is forming, it also sends gravitational waves that will be felt in one way or the other by the rest of the stories inhabiting this universe, reminding everyone that a Final Period awaits their own story, somewhere, sometime.

After having these reflections, SAM concentrated on his bench-work experiments again.

***

The stories' universe, like any other universe, was ruled by mythical Gods. Their royal court was constituted by the mighty Gods of Fate. Ever since its creation, they incorporated the Existence Dictum upon which every single story that breaks the silence of the white page rests: it will occur in a certain space within a determined period of time, and it will follow the path of destiny. This holds true in every instance, from the history of a centuries-long kingdom filling volumes, to the story bearing just a single word, of love, compassion or hate. Both stories though would have something in common: a beginning, an end and a determined fate.

Other Gods in the stories' universe pantheon did not have a major influence in the royal court's balance of power. They however, played a crucial role as they emitted their own Voices' waves which were an essential part of the ocean of human consciousness, reaching every shore. They were present in the character's brain-minds, perceived only as an echo coming from a far distance of their mental horizons, The Gods' Voices were limited to comforting those souls in times of peril, but couldn't do anything to change their luck or interact with the character's Narrating Voice, they were called the Observers.

The Observers watched the stories begin, unfold and arrive at their inevitable Final Period. They became deeply familiar with the characters´ dialogues, reasoning and emotions. In more than a sense, they had acquired a human nature. Their activity however was always looked over by the Gods of Fate; although the underlying laws governing the Final Period were unavoidable and eternal, the Observers were too close and involved in the characters' lives.

Among the Observers stood an outstanding leader and philosopher whose knowledge of the physical and the stories' universes was astonishing. They called him SAM, because he understood Science: the basic laws of physics and neuroscience underlying the simultaneous existence of both universes and whatever could not be explained through existing knowledge, he resorted to Alchemy, Metaphors, Magic, poetry, philosophical reflections etc.

SAM had long been on the royal court's black list whose Gods were inflexible, punitive and only believed in Dictums of Faith. He was different in so many ways. He started a Laboratory of Science and Alchemy that was always seen with suspicions eyes. If the rules for the stories' universe  were clearly written down, his laboratory was pointless, it should be prohibited. Other Observers could follow the miss-conceptions of the charismatic leader. SAM knew that the most difficult task a conscious being faced was to be true to himself, therefore he had to pursue his research in spite of the royal court's pressures. So he did. The last time he assisted to one of their meetings, the court saw his white gown and long white hair disappear in the distance. He left and never returned becoming a sort of an out-law. His peer Observers never stayed out of contact with him, and he continued to be their northern star.

SAM had dwelled in his laboratory for a long time, trying to unravel the secrets of the ever-present Narrating Voice in each story that so fascinated him. He used his intuition to guide him to approach the problem theoretically as well as experimentally:  Double-slit experiments, wave-function equations, electromagnetic and photoelectric essays, etc. His results made it all very clear: the Narrating Voice not only mirrored the nature of light; after coming through the eyes and triggering a complex neurological spiking activity in the brain, light *had metamorphosed itself* into an even more complex wave, the  Narrating Voice, which stands as the orchestra's director of the symphony of human consciousness.

Further research would also show him the striking similarities between the Narrating and the Observers' Voice.

****

SAM takes a break from his laboratory work and continues expressing his thoughts for himself...

—The light is an electro-magnetic radiation, consisting of these two distinct 'fields' propagating together as a single wave. The Voice also consists of such two fields, but now representing the physical and imagined reality.

Light and Voice propagate like waves, *they are everywhere* in the cosmos, just like waves in the ocean arriving at every shore. Our own Observers' Voices find place in the characters' consciousness perceived just like a far-away echo.

A universe without Light or Voice, is a universe no more.

As the Voice travels at the speed of light, it is timeless and never ages. Once generated, it will never die; every single being that harbored one, will be alive for ever in the ethereal cosmos of consciousness as a unique and unrepeatable Narrating Voice.

—Out of the different states of mind experienced by the Narrating Voice, two stand as the pillars of everybody's life: *Inspiration and Moments of Clarity*. In both instances, the Voice's waves  are required to be in a high energy-state.

*Inspiration* is difficult to define, but it is here where individual creativity has its origins. It is a most fortunate experience, to live through an inspired state of mind. Inspiration requires the simultaneous spiking of massive regions of the  associative cortex. Forming a positive feedback system in the brain, inspiration has also a direct effect on the connecting synapses of neurons in other key areas, specially the ones in deep brain centers where millions of neurons tightly packed together will play an important role in the reward-behavior, not only in humans but also in countless cases of  the animal kingdom.

When natural inspiration acts as the triggering force of such centers, you get close to the *best a human mind can be*: loving relationships of all sorts, the beauty contained in science and art manifestations; music, literature, painting, or sculpture, etc. The physical world is perceived with great optimism.

—Although initially propelled by inspiration, *Moments of Clarity* is a far less common state of mind, actually a rarity. It also has a massive concurrence of brain regions spiking simultaneously, but this time it is different. The beauty and contemplative energy of the inspired Voice, gives way to the production of light; *the brain-mind is now completely illuminated*. Reason and emotion concur harmoniously in a single space. It is of a sudden onset and generally of brief duration: it can just encompass a single moment or successive ones; during sleep or wakefulness.

The most astonishing fact, is the behavior of the story's characters with that particular state of mind. During this time, the character places the physical world on-hold, its sensing and memory brain regions forming a closed loop with the associative cortex, all of them spiking at the same time, *producing a burst of light*.

\*\*\*

Having dwelled in neuroscience, SAM had not only studied the conscious mind, but the neurological basis of the *unconscious* mind as well. He knew the crucial role played by the *emotional memory of the unconscious*; when these feelings slipped to consciousness disguised or unrecognized, the effects could be devastating if we were talking about fear, sense of guilt or preconceived notions. Only by bringing these negative feelings or rigid preconceived notions to the liberated clear mind could the subject overcome them. It meant the great distinction between free will and believing that living through certain circumstances was simply a matter of own destiny. This is exactly what happened during those *Moments of Clarity.* Destiny was set aside and free will prevailed.

All of a sudden the characters perceived a different reality altogether. For a moment they could see, hear and feel exactly their projected selves *in the future* of the story they were writing *now, in the present.* Moreover characters could watch their lives unfold, some of them even had a glimpse of how their Final Period would look like, if they continued walking the same path.

The illumination of their consciousness by a Voice in this high energy state, changed their minds. Some of them took life-changing decisions in that precise moment.

SAM undertook a study of the stories' universe cosmos, selecting a large number of *story-solar-systems* in which *Moments of Clarity* had indeed changed the characters' fate. Through his powerful telescope, he charted them in detail and noted a uniqueness in its conformation. Three peculiar *Black Celestial Bodies* were orbiting their sun in close orbit, as if they were black planets. Moreover, on the white page they appeared as *'three suspension points or points of ellipsis' replacing the Final Period;* they seemed to be preventing its formation. Intrigued by this observation, he arrived at the conclusion that these Black Celestial Bodies were in fact unique and were not black holes devouring matter, instead, they had the opposite effect and somehow provided additional fuel for the sun that illuminated the story, allowing more space and time for the story to unfold. The striking situation was that the black titan of the Final Period had been put on hold. It was obvious that this phenomenon interfered with the Existence Dictum and the story's destiny.

SAM got too involved with the stories´ fate. His sensibility, and comprehension of human nature, made him want to help the stories' characters, a drive that just grew stronger with time.

SAM's intuitive mind crystallized fundamental questions.
What was the mechanism of action of the Narrating Voice that lead to *Moments of Clarity*? Why did it occurr in some characters but not in others? Was it the same Voice or had it acquired significant changes? Could it be possible to help-generate or trigger this *Moments of Clarity* in the brain-mind of the characters? These were key issues that had to be addressed.
Every character of every story should have the right to perceive such an illuminated moment along their life-time, especially when life-changing decisions were in the making. After spending weeks in his laboratory he came to an astonishing conclusion.
The constituent matter of the Gods themselves included light, inspiration and a Voice which spread throughout the stories' universe and reached the characters' minds; but those Voices were apparently predetermined to remain passive there.

## The Discovery of a Quantum-Lock governing the Narrating Voice of the Observers

The nature of the characters' Narrating Voice or the own Observers' Voice belonged to the quantum world, where logical thinking, everyday perception of reality and common sense do not apply. After all, the study of the ethereal nature of human consciousness is complex and by no means straight forward.
SAM knew that furthering his knowledge of his own radiating Voice

was fundamental in figuring-out it's dynamics in the character's brain-mind; he would put it through a series of tests. His Laboratory of Science and Alchemy would give him the answers.

He started with the quintessential experiment in quantum mechanics: the double-slit experiment*, an essay that revealed the dual nature of light. Contradictory as it sounds, the light could take the form of radiating waves or behave just like its constituent photon particles. SAM set the required devices, a double-slit panel, and a recording screen behind. The electronics and electric supply were ready.

Standing in front of the double-slit panel, he concentrated on an intense story he had witnessed. A Voice expressed as a wave, radiated from *his self*, passed through both slits and began to be recorded on a screen behind. *As expected*, it reflected the typical stripes of an interference pattern generated by the interaction of the radiating waves generated at the slits. It also meant that the Voice as a wave lacked locality, it could not be pinpointed to be here or there but was to be found *everywhere*, reaching every corner of the cosmos and every corner of the character's brain-mind.

However, to his great surprise, the Voice's wave *suddenly collapsed* and began behaving just like if constituted by particles, hitting the recording screen only in the area behind the two slits. The striped pattern characteristic of a wave, disappeared.

Why did this happen? In order for the Voice's wave-function to collapse, it had to obey the 'measuring rule', a mathematical principle in wave functions, where a measuring device placed on one of the slits or both, would make the wave collapse; however SAM had placed no measuring sensor.

The quantum world had an explanation left: the *measuring-effect of an observer.* As bizarre as it sounds, if someone besides SAM was

---

*We advise the reader to search for the *'quantum double-slit experiment'* in YouTube-videos, where it is explained at all levels of knowledge.

*observing* the slit experiment  while the Voice passed, its wave would collapse. *Observing-eyes observing the Observers?*, or should we say *the deterministic eyes of the Existence Dictum* on behalf of the Gods of Fate?

SAM immediately perceived the existence of a *'quantum lock'* that kept the Observer's Voice confined to a single place within the character's mind. Although the Observer's Voice would be present as a wave throughout the stories' universe cosmos, when inhabiting the character's mind, the quantum-lock activated itself. SAM's collapsed Voice would only be perceived as a far-away aural echo confined to a specific region of the character's brain. This would prevent it from acting in synergy and synchronicity with the character's own Narrating Voice. He could hear their pledges, but could not respond to them or help-trigger *Moments of Clarity*.

SAM knew that he had to break the *'quantum lock'*. He tried a number of devices and procedures, but none of them worked. Then out of the blue, the concept of a kaleidoscope crossed his mind. After all, a kaleidoscope superbly described the plasticity of the brain and the ever-changing nature of consciousness. The quantum-lock targeted only one fixed image of the Observer's mind at a time, at the precise moment that the Voice was being generated; the next evolving kaleidoscope-image would then be free.
If he could reach a 'kaleidoscope state of mind' and sustain it for a short period of time, its ever-changing image would make it impossible for the quantum-lock to take place. He finally achieved it *through meditation*. Concentrating in a particular story's character and  envisioning the *Moments of Clarity* that would constitute a helping hand at the precise moment and place  where they were needed; the rapid succession of images made it impossible for the quantum-lock to set in. The wave held and did not collapse.

Two electromagnetic waves traveling in the same space and direction add together forming *a single wave* of higher amplitude.

An Observer's Voice reaching out to every corner of the human mind would have also have an important *synergy and synchronicity* with the character's own Narrative Voice. The resulting *synergic wave* would help bring to the conscious mind, negative feelings or experiences entrenched and hidden in the unconscious, revealing their true ugliness and destructive power; an event made possible through *Moments of Clarity.* Enslaving chains could then be broken. SAM was ready to test it for the first time.

During his daily routine as an Observer, he perceived the ongoing story of a teenager. His telescope showed her *story-solar-system* with no orbiting Black Celestial Bodies. She was living through a tough tipping-point in her life, and had to choose between two different opposing paths. She already was configuring a mental logical workout that justified a series of wrong decisions. It was morning and she was walking alone in a park to meet someone. SAM knew that this moment of solitude was ideal to bring about a change in her. His Voice could be sensed as an aural echo in the back of her mind, as a regular Observer. Then SAM concentrated in a kaleidoscope state of mind and his Voice's wave came into synergy with her own inspiration lighting up her mind. In a *Moment of Clarity,* she could see *her self* from an outer perspective; the chains of addiction and fear of a totally misunderstood relationship became evident to her, so was the dead end she was aiming to; she stopped walking. A few minutes later she saw him coming in the distance but now her state of mind was so different, she recognized in him a monster that was dragging her to the bottom of deep waters. How could she have missed it! She turned around and started running. The wind hitting her face drying her tears, tears of relief; she would scream and ask for help if he tried to catch her. Never before she felt so free.

Back in the Laboratory of Alchemy, SAM found himself lying on the floor slowly regaining forces that allowed him to stand on his feet again. He realized by now that every time he avoided the quantum-lock during the kaleidoscope state of mind, a fraction of his mass would be transformed into the needed energy's Voice's wave that could help the character achieve *Moments of Clarity*. The Gods of Fate had placed a prize for the Observer that dared to change the story's fate. They would pay with a piece of their own existence.

Once recovered, he couldn't wait to observe if there were any changes in the woman's *story solar system*. He looked through his powerful telescope and now found three Black Celestial Bodies orbiting its sun; the story's fate had changed for the better, her soul flying again as an unharmed bird.
A smile lit his face.

SAM had important stories in mind where he wanted to replace the Final Period with  Black Celestial Bodies. He applied himself to the task and the results were astonishing. The rest of his fellow Observers had to know about this. He called for a general summit to be held at their usual hide-out: The Cosmic Theater.

## THE SUMMIT OF THE OBSERVERS
### The Cosmic Theater

Observers came together from time to time, to share their experiences at their usual hide-out of the stories' universe. They had devised an extraordinary way to do it; the white page had evolved into a virtual huge screen *where the actual stories could be projected*. Image and sound came from the Voice's waves of the designated speaker. A witnessed-story stored in the memory of just one Observer, could now be shared by the whole auditorium. It was a simultaneous and enriching experience for everyone.
This Summit Meeting was carried out in secrecy. The mystery surrounding the meeting and the fact that SAM was the highlight speaker and presenter, created great expectations among the Observers.  SAM had always been their most respected leader,

truly a northern star.

SAM stands in front of a huge Observers' audience. He begins delivering his talk.

—I thank you all from the bottom of my heart for being here. I had to inform you that my Laboratory of Science and Alchemy has lately produced astonishing findings regarding the properties of our Observer's Voices which bear important implications for everybody.

—As Observers of the stories' universe, we have witnessed together millions of stories, how the dust of human consciousness begins forming a sun, around which an orbiting planet will harbor a story to be written; how it starts, unfolds and comes to its Final Period, a Black Titan that will keep the whole story, its characters and space-time densely held and packed together within a confined tiny black space by the powerful Forces of Fate.

—The presence of our Voice in the character's minds is perceived as an aural far away echo and can't do anything further. In a sense this is right, for in essence we are just that, Observers. Without entering into the mathematics and physics of wave function equations, I have discovered an in-built 'quantum lock' present in our Voice that operates only when we are inhabiting the character's consciousness. It is deterministic, collapsing and confining our Voice to just a single location preventing them to interact with the character's own Narrating Voice.

—I am here to report to you, that I have discovered how to avoid our Voices' wave collapse, through what I have called 'a kaleidoscope state of mind' achieved through meditation. It works. Our Voice's wave will act in synergy and synchronicity with the character's own Narrating Voice's waves. This allows us to be confluent with the inspiration already present in their minds, helping story characters in crucial moments of their lives to create enlightening *Moments of Clarity*. The implications are enormous.

—The characters experiencing these *Moments of Clarity*, will be able to change his or her predetermined destiny for a better future. This in turn will reflect in their *story-solar-system*: three peculiar Black Celestial Bodies will be found orbiting around its sun. They are not Black Holes, on the contrary, they do provide the story's sun with additional energy, space and time. In more than a sense, they act as *suspension points replacing the Final Period.*

—My findings do not stop here. Most remarkable is that stories finished long time ago, can also override their Final Period by constituting themselves into *parallel-solar-systems*. We have to remember that the character's Narrating Voice once created, is of a timeless nature and now forms part of the wholeness of the stories' universe.

In some instances, we can actually go back in time and contribute to light those brain-minds with *Moments of Clarity*. In other instances, we can resort to alchemy procedures to bring elements of reality together that are or were already present, helping them crystallize into a new promising future.
In both cases we will witness the formation of another *story-solar-system,* where the main characters will write a *different story altogether*.

As you can see, fellow Observers, this actually contemplates a new reality within the stories' universe, redefined by the quintessential essence of human consciousness: imagination. Without imagination the stories' universe would simply not exist; but Black Celestial Bodies mean *imagining a better world.*

—My experience so far has been highly rewarded. I have seen birds flying out of their cages to experience freedom of choice, freedom of pain, the promise of a daily hope, a vast blue sky right in front of them. There is a catch however. The process of maintaining the wave properties of our Voices, even for a limited time, comes to a price. Every time I avoided the quantum-lock preventing my Voice from collapsing and adding-up with the character's own inspiration was made at the expense of converting a fraction of my own mass to energy. Apparently this is the price we Observers would have to pay, with a piece of our own existence.

The audience kept silence as he finished his explanatory speech. Only the Observers' eyes reflected the implications of such a discovery.

SAM continued his presentation,
— I will be projecting four stories, each of the stories will be shown as it happened, and then as it became a Black Celestial Bodies' story. Important comments and observations will arise among the audience; I will ask you for your patience till my presentation is over. Thereafter I would like to have your feedback and answer any questions you might have. Finally we will have a session on how to achieve the required meditation aiming at the kaleidoscope's state of mind that will allow you to break the Voice's quantum-lock.

***

## THE ROSE IN THE DESERT*
Santiago de Chile. Pinochet's Dictatorship *circa* 1975

—I will begin my presentation by projecting an introduction to the first story, said SAM.

After the military coup d'état to the socialist government of Salvador Allende in 1973 in Chile, Augusto Pinochet was installed as prime minister. What followed was 17 years of a dictatorship where violence and persecution were commonplace; thousands of civilians were detained and many of them murdered. The discovery of mass-graves in recent times**, revealed the magnitude of the atrocious nature of the dictatorship. Some of the mass graves were located in the Atacama desert where a concentration camp took place. Among the bodies found there some were incredibly well preserved after decades of being buried; they belonged to young women whose beauty still could be recognized in their faces, their delicate hands brutally tied with ropes before their execution.

The magnificent starred heavens of the Atacama desert, were just waiting for an Observer's imagination to ink-skate on the white pages of the stories' universe, *writing a different parallel story* about the victims' destiny. It also stands in memory of defenseless innocent people that have violently lost their lives anywhere in the world, regardless of age, gender, religion, creed or race.
Follow me with your imagination.

A middle-class Chilean family is sitting at their dining table having their evening meal together; parents and their only daughter. The conversation starts as usual on Pinochet's military dictatorship, that gave a coup d'état to the socialist government of Salvador Allende two years ago. She is a medical student and so is her boyfriend. Both are active in elaborating underground flyers and printouts; they

*Original story conceived and written by the present author S.Cordero
**Documentary of 'Nostalgia for the Light' by Patricio Guzmán.

are part of the insurgent socialist movement. Tonight, however, their situation is deemed as 'critical' by both parents; they have learned of recent raids and detention of people they knew. She could well be on the list of future detainees, the socialist movement had been infiltrated by young government spies. Parents want her immediately out of the country and have come-up with a plan to do so, they already have gathered the needed money for the escape through the Peruvian border. She regards fleeing as treason to all the martyrs that already gave their lives for socialism; further, she cannot leave her boyfriend behind. Her parents insist, there is no time to loose. She will think it over and discuss it with her boyfriend; she will have a response within days.

During sleep they are raided. Four armed men storm the house, she is on their list; the noise of the front door lock being violated wakes them up. Parents will face the commando and she goes into hiding. They ask for her, parents reply she stayed in a friend's home due to her exams. As they search, books and flyers reveal her link to the socialist movement. The commando's officer takes his gun out and says he will count until 10 and shoot both of them if they don't reveal her whereabouts. He begins the countdown in a loud voice; the daughter was able to hear the conversation and comes out of her hideout. Parents beg the commando not to take her, they offer themselves as a substitute. Useless, she is taken prisoner. Parents lie on the floor devastated and desolated.

She is put into a truck full of beaten up prisoners, they go the army´s headquarters. At arrival she finds out her boyfriend is also there, someone betrayed them, they even got photographs showing them handing-out flyers. All prisoners are identified and sorted.

The least fortunate, considered as a menace to the country's stability, are taken to the Atacama Detention Center where they are going to be killed. Boy and girlfriend included.

After a long and rough ride, they arrive at the Atacama Detention Center and are concentrated in an open area. An irony of life, the boyfriend had been in this place before as a teenager; it was an abandoned factory. They camped with friends close to this facility to watch the starred heaven, they even visited an observatory the Germans had built in that area. The detention center and the observatory were relatively close, within half-a-day walking distance. He was positive that the observatory was located north. They had to escape, reach the observatory and ask the Germans for asylum. It was their only chance of survival. As the afternoon slowly drifted into the night, the beautiful Atacama starred sky began appearing. He finger-pointed the northern star to her that had to be followed.

The medical student was correct about the German observatory, where a young German astronomer, a student working for his phD, had decided to take a break; he had talked with his advisor before spending a couple of nights alone under the Atacama desert starred sky. He wanted to experience a personal dialogue between him and the stars in the solitude of the desert. With back-pack and motorcycle, he set campfire 5km on the outskirts from the observatory.

Back in the prisoner's camp, both manage to get close to the entrance. Continued arrival of trucks and the dim afternoon light causes some confusion. Hiding behind the moving trucks, they approach the exit. Just when they are about to cross it, a guard directs the reflector lights on them. He raises his hands diverting the guard's attention while telling her to keep walking pretending he was alone. It is heartbreaking, she escapes and remains outside close to the perimeter wall, while the main door is being closed. He had sacrificed himself for her.

Now she is alone, depending entirely on her survival instincts. Drying the tears in her eyes, she keeps walking following the direction given by the northern star. During her long walk, she stumbles and falls, her protective right hand avoids a direct hit of her face with the ground. Still lying on the floor, while recovering her breath, an unexpected finding catches her attention. Right before her eyes, a flower has inexplicably grown in the middle of the desert. It is a beautiful white rose that shines under the light of the stars. She is taking it with her and places it in her pocket.

After having walked all night, exhausted and dying of thirst, shortly before dawn-break she spots a campfire, it is the German's scientist one. The hum of the jeep motors is now to be heard in the distance, they are after her. He is preparing coffee and just placed the pot of water over the fire unaware of the unfolding moments. She can see him, but still 200 meters away and too tired to run; she is not going to make it and shouts with the last energy she's got left,
— Help me! Help me please, they want to kill me!

The astronomer hears her screams in the distance; stunned at first, he gets her desperate situation. She must be a runaway, the military jeeps want to capture her. His reflexes react and set him in motion, he reaches for the motorcycle keys in his jacket's pocket, gets on his motorcycle turns the engine on and goes to meet her.

He arrives. She is exhausted and can barely speak.

— Get on! Get on! Hold me tight and don't fall down!
The German astronomer says, as he heads for the observatory at full speed. A 5km chase begins. The jeeps overrun his camp.

The soldiers begin firing shots, the sound of the bullets hitting nearby make him maneuver and not travel in a straight line; 50m away from the observatory, they have almost made it. However he is seriously wounded in a leg by a shot. They fall down. Bruised but unhurt, she quickly recovers, stands up and wants to help him.

—No! He finger-points to the building.
Go inside!... Germany!... Tell them to call Germany!

She makes it inside the observatory.
The soldiers arrive and are short of shooting the astronomer.
He has been seriously wounded, is bleeding badly and shouts,

—German, I am a German citizen, *ciudadano alemán* .

The government and military officials of both countries communicate over the radio, and have come to a deal. He will be airlifted, but the escapee, deemed as a terrorist, must surrender.

The German takes his shirt off to self-apply a tourniquet to stop the bleeding. The helicopter arrives, paramedics want to bring the astronomer on board. Soldiers refuse. If not treated promptly he will die. Negotiating parties talk again. She must come out, that is the deal, take it or leave it. Sharp-shooter soldiers are posted, they have orders to bring her down as soon as she  walks out of the building. The building door opens. The prisoner can barely be seen standing alone at the entrance. The astronomer is now allowed to be placed inside the helicopter. As the escapee walks into open space, an unexpected group of four scientist surround her forming a protective shield with their arms held together. They come from diverse countries as shown by their T-shirts. They walk her out to the helicopter risking their own lives. She must get on board.

The deal was broken. The sharp-shooters have their nervous fingers twitching on the trigger, they didn't count on this and can't fire. She makes it inside the aircraft. The helicopter doors close. A bazooka now points to the helicopter as it takes-off. At the very last second, the officer stops the bazooka-shooter and lets the aircraft go.

Inside the helicopter the astronomer although conscious, is dying due to the hemorrhage; his blood pressure has dropped to 60/40.

# The Story of Stories

The assisting paramedic gets close to his ear and tells him:
—You have bled too much, you need a transfusion, urgently, right now! What's your type of blood?

—A+, he answers in a low voice.

—We need a blood donor! Anyone in this aircraft that is A+?— Asks the paramedic twice in a loud voice barely to be heard due to the engine's noise.

—Me! I am A+, totally sure; I am a med student, she answers.

While the paramedic is drawing blood from one of her arms, astronomer and escapee look at each other. He is whispering some words and she needs to get close to be able to listen,
—Thank you... he says

—You saved my life and risked yours, she replies.

—That's a beautiful flower you are carrying in your pocket, he remarks, barely moving his lips.

—Yes, I found it in the middle of the desert after escaping and walking all night before I met you; I guess it is a gift from the stars.

The astronomer replies with an almost imperceptible smile,
—... so are you...a gift from the stars...
Thereafter he closes his eyes. Has he collapsed?

She takes his hand and tells him with tears in her eyes,
—Hang on...hang on...please, don't leave me...

***

## EL QUIJOTE de la MANCHA

The audience remained in silence after the first Black Celestial Bodies' parallel story had been projected. They immediately sensed they were witnessing the birth of a new era in the stories' universe. SAM was now aiming at the very pillars of literature.

—What follows is an all-time high of the Spanish literature, El Quijote de la Mancha by Miguel de Cervantes.

Next, we'll see what imagination can do for such a brave Knight.

—The quintessential Knight of Idealism, El Quijote de la Mancha, lies in his deathbed after a life-long commitment of fighting for a world of moral and human values invisible to others. A beautiful story full of metaphors, takes a strange twist because now, at the very end of his life, lying sick and tired he surrenders to a brutal realism. He denies his *idealistic self* and names his past actions "acts of lunacy". By this, in just one act of despair, he surrenders each and every one of his hardly won battles and conquests achieved during decades of his knight quest. During that period of time, he pursued his imagined reality, always being confronted by his squire Sancho Panza, who's view of the world was without sophistication whatsoever; couldn't conceive a metaphor or attribute of any object or person except for the realistic context presented to his uneducated mind.

Two opposing realities riding together on the Spanish plains of La Mancha, as dissimilar as their mounts: The Knight of El Quijote riding on Rocinante, his tall and magnificent mount that always brought him closer to the heavens –an old and weak horse for Sancho– and his accompanying assistant riding at his side on a stubborn donkey.

Don Quijote's imaginary vision of the world can be depicted in two passages:

1. The heroic Quijote's battle charging with his mount, in a medieval-just manner, against a group of menacing giants that could devastate nearby villages. This left him severely wounded lying on the ground. (Not giants at all, but just windmills to Sancho).
2. The conquest of Dulcinea del Toboso, whose beauty captured Don Quijote's heart, becoming his Queen and Lady. She needed to be liberated from a malicious spell that was enslaving her. (Just a woman of rude and vulgar appearance to Sancho).

All his endeavors being reduced now to nothing by the hammering of brutal reality upon them, its pieces falling into the abyss of the Final Period and held in confinement by the powerful Forces of Fate for eternity.

All Don Quijote needed was a helping Voice within his mind to bring about a *Moment of Clarity* to his consciousness. Let the alchemy of our imagination crystallize this *parallel story* on the white pages of the stories' universe...

Don Quijote lies quietly on his bed. He feels exhausted and abandoned by inspiration. He cannot build imagined reality out of the physical world anymore, dragging him into a deep depression. He surrenders his dreams and even his name as The Knight Don Quijote de la Mancha; with a fragmented existence, he tries in an act of despair to reconstitute himself under his past common name of Alonso Quijano, in the very final hours of his life.

*Suddenly*, an unexpected burst of beautiful fireworks breaks into the somber night. A *Moment of Clarity has lighted his mind* and brought him to his *very self* again, where he can perceive what he is and was meant to be: The Knight of Idealism. Metaphors start emerging and an urge to stand up from the bed, begins feeding the muscles of his limbs with energy. His arms feel stronger, his legs and feet ready to go.
—Sancho! Sancho! Where are you! Come and hurry, we have a lot

to achieve! Hand me all of my armor!

He stands up, looks formidable, with a focused vision. He reaches for parts of his armor and begins fitting in. Sancho is absolutely bewildered. He was feeling sorry for his Lord who was going through agony after hitting the reality-wall head-on; but now suddenly, he regained strength and determination like never before. He helps him dress in full armor.

The Knight of Idealism is standing tall, holding a sword in his right hand high-up, he pronounces:

—Iron of justice, my eternal friend, help me fight for the weak, for the oppressed, for an imagined reality that is in great need for a defender that will defeat the ghosts and demons that afflict it.

Sancho is in a complete state of surprise,

—My Lord, I could encourage you to keep on flying in the heavens of idealism, but I would be misleading you. Sooner or later you will have to land on reality, which means going through disappointment and suffering again. Your old age and health don't give you much more time to settle things down. Remember that reality makes always the final statement.

Don Quijote,

—Reality. What is reality Sancho? The perception of the physical world through our senses arrives as images and sounds in our head, but it is our mind that puts them together in the form of thoughts; these thoughts surface in our consciousness and it is precisely here where they acquire a meaning; the shape and color of this meaning will largely depend on what your mind already knows. In other words Sancho, *you can only see what you know* and even more important, *you can only see what you feel.*

There is no greater truth than 'beauty lies in the eyes of the beholder' for the face, hair and body of a woman turn into a cascade of emotions within our mind Sancho; it is from within your mind that you should begin looking at the world.

—I was meant to be the Knight of Idealism that will help redefine reality. Sancho! Prepare our mounts!

Sancho,
—But my Lord, it is dark and cold outside, it is night-time!

Don Quijote,
—Precisely Sancho, precisely. It is in the dark where the muggers and phantoms hide. Let them know their felonies will be stopped and a new day for mankind will arise at dawn; Don Quijote de la Mancha is back!

Don Quijote is exiting his room. Replacing him on his bed lies now Alonso Quijano, a broken and defeated man. Reduced to a motionless body, Alonso's sad eyes are living their last moments of light, just waiting for the relieving death's fingers to close them.

Don Quijote stops for a moment, turns around and looking at the dying man, directly into his eyes, says:
—You have given up, renouncing to everything you fighted for, in the last moments of your life. Where is the stamina that feeds your soul to keep on fighting, that provides a healing salve for the wounds inflicted by reality? There is no way that I could die as Alonso Quijano; instead, this deathbed belongs to you now, a gesture you might even appreciate. Where have you left your dignity and your love for life? Imagined reality has abandoned your consciousness and surrendered to a brutal realism. But not me, El Quijote de la Mancha, The Knight of Idealism, *will never lay down and die*. I am the air filling the lungs of true warriors, I got to be there for them. Always.

—Today Sancho, we will ride into the night and with our sword strike a tear to its dark and somber vault, to let the stars in the heavens light up our dreams…

***

## ROMEO AND JULIET

SAM made no pause for his next projection, he continues,
—Now let's visit the story of the two most celebrated lovers of the stories´ universe, Romeo and Juliet by William Shakespeare. Unsurpassed love and beauty were marked with the fate of destiny; a series of tragic events would corner the lovers and ultimately claim their lives.
SAM added with a sense of conviction,

—Characters in a story become truly independent beings from the writer's mind that created them; thereafter, our Observer's imagination can give them an alternative to their Final Period. This is how Romeo's and Juliet *imagined parallel* story unfolds...

Romeo enters the Capulet's crypt where Juliet lies. The news of her death were devastating. However, she had been brought here in a state of temporary narcosis due to a potion she intentionally drank; simulating death  was  part of a plan to avoid her father's orders to marry someone else and have a chance to escape with her Romeo; circumstances however had another plan for her.
Not being enough anguish the burden of Juliet's 'death', Romeo continues adding fatalities; he encounters Paris in the crypt, who also came to mourn Juliet. Believing Romeo was a vandal, Paris confronts him but loses his life in the battle.
After having killed Paris, all Romeo wants to do is to leave this catastrophic world to reunite himself with Juliet. He approaches Juliet's body. He touches her face and feels it is still warm.

—Oh my beautiful and beloved Juliet! Look at you, a flower lying in this disgraceful place! This is the saddest day of my existence; fortunately I brought with me this small flask filled with venom. It is my best friend now and shall lead me to your side, wherever you are. Do not despair my Juliet, my death will not take long, but before I drink it, I will kiss you for a last time in this world and will continue to do so in heaven.

Romeo closes his eyes and barely touches Juliet's lips...he can't leave them, not while they are still warm. Suddenly he perceives something like thin air coming out from her nose. How can this be? Is she breathing? Romeo has a *moment of clarity.* Juliet might not be dead yet! He wets his lips and gets close to her lips and nose again, this time concentrating on the perception of any sign of breathing; a fine stream of air coming out of her lungs dries his lips.

—Yes! She is breathing!. He raises his fisted festive hands in the air and says it loud again: Yes! She is still breathing! My Juliet you are still alive!

He looks around at the crypt where royal clothes hold decomposing bodies over cold marble, and says,

—You don't belong in here my Juliet, for your heart and mind are full of life. I know all of these bodies and unfortunate circumstances surrounding you, want your shining beauty to light up their disgrace. They already had welcomed you, *but this story is changing its fate right now!*

Don't get me wrong, I do respect you all, nobles of royal descent, and have no prejudice or hard feelings against you. I can only leave you my prayers, for Juliet is coming with me.

Let history put the blame on my name for doing this. History needs you to remain in here my Juliet, but this Romeo needs you even more; and while I got two arms to carry you out of this crypt, history has no arms to stop me.

A highly inspired Romeo slowly puts his hands between her body and the marble; he feels the warmth of Juliet's body on the palm of his hands, contrasted with the coldness of the stone. His arms are now beneath her body; finally, he lifts Juliet and separates her from her tragic fate…

***

## SOR JUANA INÉS DE LA CRUZ

After Romeo and Juliet projection, SAM made a pause and directed himself to the audience.
—The last story to be presented deals with persecuted scholars due to their scientific inclinations. These freedom fighters, deserve a special place in our minds. Such is the case of Juana Inés a catholic nun in Mexico* who became one of the most celebrated literary authors of the 17th century.

In her flagship poem *'First I Dream'* (Primero Sueño)** she redefined reality and placed consciousness in confluency with nature and the universe through a language constituted by science and poetry.

—To better understand Juana Inés, I have built a metaphor around her story that will allow you to envision what she actually lived through. Imagine yourself in a community of caterpillars, of a species that does not metamorphose into a butterfly. Every single citizen does  what a caterpillar is supposed to do; and there is no conflict whatsoever. One fine day, an outstanding female inexplicably has the need to construct a cocoon. She metamorphoses into a butterfly.

Her butterfly state gives her the chance to see the universe from another perspective. She flies high and everywhere: poetry, literature, mathematics, physics, even music; she explores them all. Her published achievements are such that the ecclesiastic status-quo feels eclipsed and threatened; they finally halt her flying. She is cornered and intellectually executed. Her wings are cut off.

---

*Mexico was a Spanish colony until 1810 known as the 'Virreinato de la Nueva España'. To simplify, the name of México was used instead.
*Additional information on Sor Juana's referenced text on page 64.
**Poem first published under this title but also known as *'The Dream'* (El Sueño).

SAM's projection began.

—Sor Juana's life did have a head start. At age 3 she is taught how to read, at age 8 lives with her grandfather and has access to his personal library; her beauty and intellectual charm brought her as a protégée in the Viceroy's wife court. At 18 she was a radiant woman that did not go unnoticed by the court's men. For a woman of her talent and intellectual ambitions, marriage was not an option. Women at that time were generally treated like property, they had no voice of their own. They had to endure childbearing of numerous children, alcoholism and abusive husbands. She decided to become a nun.

The ideal convent would be waiting for her, the Jeromite in Mexico City, which she would never leave. She continued having the Viceroy's court protection and the convent allowed her to pursue her intellectual ambitions. The butterfly was flying high and exploring the universe. She could devote entire days to study all the Greek and Roman classics that crossed her way in convents and monasteries and even be in correspondence with intellectuals in Europe. Her work thrived, she published plays, poems and prose. Her poem, *'First, I Dream'* (Primero Sueño) makes reference precisely to what the human mind was meant to be, to explore every area of knowledge, from physics, astronomy to music. Nothing could get in the way to stop the intellectual drive of the human consciousness to explore and understand nature and the universe.

These statements did not go unnoticed by the ecclesiastic hierarchy. Up to now, they had tolerated her achievements, but she was eclipsing their image and her way of thinking was incompatible with the understanding of the universe through religious believes. Scholars with scientific inclinations were persecuted by the Inquisition. Any excuse would be used to put her away.

Her decline began when she was caught in a dispute of power

between two prominent bishops, Fernández* and Aguiar*, fighting for the highest ecclesiastic position that was vacant in the Spanish colony; it could eventually lead to the granting of the Viceroy's post. Apparently Sor Juana was asked by Fernández to philosophically question Aguiar's doctrinal stronghold,  Antonio Vieyra, a well-known theologist from Portugal. She did so in an extensive and elaborated  published letter. However in spite of Sor Juana's scholarly argumentation, politics favored Aguiar.

Not only did Fernández retreat, but changed sides; he wrote her a compelling letter 'strongly suggesting' to re-orient her talent to the praise of God as conceived by the church, and to stop dwelling into scientific and philosophical mundane issues to which she already 'had dedicated a great deal of time'; every member of the church had to submit to obedience. Bishop Fernández, even hid under the pseudonym of a nun (Sor Filotea de la Cruz) avoiding to use his own name.

Sor Juana was left standing in the middle of the nowhere. In an extensive letter defending her case, Sor Juana names Hypathia of Alexandria, the first woman-mathematician of antiquity and martyr, who was murdered by a group of religious fanatics that opposed scientific pursuit.

An adverse domino-effect soon followed; the Viceroy's court protected her no more, and Aguiar, the highly-feared bishop, stepped in. He attacked Sor Juana openly; she felt all alone, facing a pack of inquisitorial wolves with no helping hand in sight. In a gruesome gesture of destiny, her last bastion and dear confessor Miranda*, became the executioner. He was the one in charge of cutting the butterfly's wings.

Just like the prisoner facing the executioner, she is stripped from all of her belongings that are her academic and intellectual possessions accumulated throughout her lifetime: a whole library of precious and diverse books, devices to study astronomy, physics, musical instruments, and even her writings; they will all be confiscated and sold for the 'benefit of the poor'.

The humiliation went much further, she also had to renounce to any further mundane intellectual activity and stick within the boundaries of religious and ecclesiastic praise.

Without wings, the butterfly couldn't survive. Four years later, after being confined to the cell of deep depression, one of her last written pages bears her signature signed not with ink, but in red with her own blood. Months later, she dies.

Was there someone in the 17th century world that could have praised her as a beautiful woman and at the same time as an authoritative scholar? The answer is surprisingly yes! (1). He was actually an intellectual mirror-image of her and just about the same age.

Being an accomplished poet and exploring scientific endeavors, Francisco Alvarez from Colombia** came across Sor Juana's published work: prose, poems, plays, and he immediately fell in love with her. Being a lone and rich widower, he brought her to his imagined reality under the name of Nise, a name holding the same four letters as Ines. He couldn't use her name for she was a well-known nun; love and admiration led him to write beautiful poems to Nise.

The poets lived along parallel lines that never crossed each other. Francisco wrote an extensive letter to Sor Juana dated in 1698, but Sor Juana had died in deep depression 3 years before in 1695.

As the *original story's projection finished*, SAM directed himself to the audience,

—Fellow Observers, you can clearly see how the elements of reality were already present  that would have allowed the butterfly to

---

(1) Please go to page 64 for historic reference.

** Today's Colombia was a Spanish colony until 1811, called 'Virreinato de Nueva Granada'. To simplify, the name of Colombia was used instead.

keep on flying for the benefit of mankind. Let the alchemy of our imagination crystallize a *parallel story* in the stories' universe.

—Travel with me back in time. Francisco must have been around 41 years old and Sor Juana 37 when the following crucial night unfolded. Let the projection on the white screen talk for itself.

Alone in her convent cell, surrounded by the darkest night, the flicker of a candlelight being her only companion in such grave moments, Sor Juana is answering Sor Filotea's letter, a letter that brought with it the hammer of imposed obedience that every single member of the church had to follow. She knew that the nun's name of Sor Filotea was used by bishop Fernández in previous communications of this kind; but it was not nearly as dangerous as the mighty bishop Aguiar who now was attacking her openly. In addition, her relationship with her dear confessor Miranda came to a halt. Cruel irony, Miranda also served as a doctrinal advisor to the Inquisition. To complicate things even further, she did not count on the Viceroy's support anymore. She was cornered with no way out.

Sor Juana is simply devastated, although she is defending her case, she will be facing her own intellectual execution, renouncing to the purpose of her life, condemned to a monastic and repetitive daily life that had no room for imagination.

How many pages did she fill with uninterrupted hours of handwriting, under the same candlelight, giving way to an imagined reality so far from the convent's routine she was living in? It was the white page universe that became her best friend where Sor Juana could turn into *her self*, Juana Inés the beautiful and passionate woman, Juana Inés the scholar and academician.

She stopped writing, brought her hands to her face and cried; then, she began looking into her drawers and cabinets, almost as a good-bye gesture to her personal documents; the tears in her eyes hardly let her see them. Half buried among papers lay an envelope that she still had to forward to the right person. How could she have forgotten! She herself had attended a knocking on the front door while walking on the corridor leading to the chapel, and received it about a week ago.

The mailman delivering the letter handed the envelope to her. It had been sealed by the Colombian Postal Office in Bogota-City. It lacked a sender's ID and was directed to a person bearing a single name.

*Please forward this letter to Nise*
*The Jeromite Convent, Ciudad de México*

Sor Juana was puzzled and played for a moment with the envelope in her hands,
—There is no one here that goes by that name, she told the mailman.

The deliverer kindly insisted, while finger-pointing the written address on the envelope,
—Here it says The Jeromite Convent, and that place is right here... please take it, it is very important; I can only hope you will be able to find the right person.

Hesitating, she promised to find out who that person was, but with so many crucial things happening at the same time in her life, she simply misplaced it among her papers. Now that she was going to hand all of her writings to the head-nun, she would include this envelope too.

Sor Juana had finished writing. She now would fall into the abyss of the Final Period for eternity that contained the powerful Forces of Fate. Her flagship poem 'First I Dream', was over; she was awakening to a nightmare.
With tears in her eyes, feeling like having fought a one-soldier battle all her life long, and defeated in the end, Sor Juana is ready to

dismiss her eternal friend, the candlelight's flicker. As she inclines her head to blow over the candle, the letter to be forwarded to Nise will be the last image to go into darkness.

She stops for a second... takes the envelope in her hands and reads the words written on it a last time... Nise?... four letters... same four letters as in Inés... how come that person doesn't have a second name? Is Nise a code name... a code name for Inés? Is it meant for me? Is it someone trying to communicate with me and this was the only way to do it?

Sor Juana decides to open the letter. The candlelight is still lighting the top of her desk. She carefully opens the envelope and takes the letter out. The name of Francisco Alvarez appears for the first time before her eyes.


*Dearest Nise,*

*I can only hope that the hands holding this letter and the eyes reading through it belong to you, Juana Inés. You are living within a faith's fortress, surrounded by thick guarded walls and I just didn't find another way to convey a message to you.*

*I have read and studied every published work you have done, I imagine your delicate hand gliding on the white page expressing poems, plays, scientific thoughts and even music; while your beautiful eyes guide the pen that inks them for eternity.*

*I lead a personal life in solitude, for my wife passed away some years ago. May I emphasize the word solitude, abysmally different from loneliness. A solitude you and I share, air beneath our wings that keeps us flying in the imagination's skies and helps us see the universe from another perspective.*

*I do dare to tell the literary genius, that I also write poems, only because these poems are dedicated to you, Nise.*

*Earthly material things I own, properties, plantations, real estate, gold. However I feel so different from my peers here in Colombia. I have been reluctant to marry another woman, there is not much that we could share.*

*How can I possibly overlook Nise's existence? There must be at least one space and time, in the stories of our lives, where we can stand facing each other. I would give everything I've got for that moment to happen.*

*Now that you know of me, please answer my letter. Even if you decline my friendship, that page bearing your writing will be enough to keep me inspired for the rest of my life.*

*Every night, before going to sleep, I always read a piece of your extensive poem 'First I Dream' hoping that through my dreams I can escape to another reality, where you and I could live and write poetry together.*

*A poet's heart and soul, yours forever,*

*Francisco Álvarez*
*Santa Fé de Bogotá, Colombia*

Sor Juana is stunned. Does this mean a last-minute salvation door opening for her? Who was this Colombian poet she never heard of, writing a love letter to her? Not understanding the ensuing circumstances, she clings to the letter like a castaway clings to the only floating log left from the shipwreck. She wastes no time and answers the letter immediately.

Francisco Álvarez, Poet.
Santa Fé de Bogotá, Colombia.

This is the first time ever, I have received a letter from a man declaring his love for me; it had to be said through the words of a poet. Emotion and reason have always walked together in my mind, never before did my emotions travel through rapid rivers.

You talk about solitude and loneliness. Well Francisco, this is a very lonely moment I am living through. High-ranking bishops have the perfect excuse to put me away. As you know, in my 'First I Dream' poem, I praise the human mind God has given us, for it bears a consciousness that will explore the universe, unraveling its mysteries and arriving at the physical laws that govern it. Like Hypathia of Alexandria, the first mathematician woman in antiquity, I am facing a pack of wolves. Their predeterministic theological thinking want to chain my mind to rigorous Dictums of Faith.
All of my belongings, my book collection, my scientific devices and my musical instruments will be confiscated and sold. Moreover I will have to publicly renounce to my thinking and writings. Thereafter I will have to submit to obedience and lead an exemplary nun's life for others to follow. My heart is full of fear. Will the inquisitorial process stop here?

Just as I was going to dismiss the candlelight's flame and be surrounded by the darkest night, I opened the letter sent to a Nise I didn't know before.

I ignore the succession of events that will unfold here for me. Nevertheless, I want to thank you for your kind letter that has brought air into my lungs and hope to my mind.

Sincerely, Sor Juana Inés de la Cruz
The Jeromite Convent, Ciudad de México

Sor Juana must send Francisco her answering letter at once; she has friends in the mail service that will assist her.

Twenty days later, Francisco Alvarez is reading her response in Colombia. He can´t believe his eyes. That very day his astonished servants at home, pack his suit cases for an imminent trip to Mexico. His right-hand collaborator is coming too. He goes to his safety box, he is bringing with him gold-coins, not only enough to pay for the expenses of the trip, but enough to buy real estate; now he has become a book and art-collector. There is no time to loose.
—Prepare the carriages! Mexico is waiting for me!— He orders.

The day came when Francisco Alvarez knocked at the convent´s doors. He presented himself as a book and art collector who was willing to top any quoted price in gold for Sor Juana´s intellectual possessions. The set price was high, but Don Francisco could afford it. The only condition on behalf of Don Francisco written in the contract, was that Sor Juana should put her name in handwriting on every possible object or document to valid its authenticity. He personally would supervise it.
A church official closed the deal with him and early next day both signed the contract in his office, which was confluent with an ample room where Sor Juana's belongings were kept under a locked door. The official received the heavy sack full of gold-coins, stood up and placed it in a safe-box. While doing this he says:

—You have to understand...Don Francisco, that it is not greed governing this transaction, but our enormous drive to help the needy and poor. Sor Juana will be here in a brief moment to begin signing her work. Please understand the sacrifice she is making in every regard.
From now on, her talent will be devoted to the praise of religious faith, a role model for other nuns to follow.

The church official now invited Don Francisco to enter the room where Sor Juana's belongings were kept. He opened the locked room revealing its precious content; he said with a convinced tone of voice:

—Don Francisco you just have acquired a magnificent and unique book and art collection, congratulations. According to the signed contract, all what you see in this room, is now officially yours. I do have to attend other important issues. Sor Juana will be shortly here. Be patient if she breaks down and cries, it is expected, you know the fragile nature of women...give her a minute or two to regain her pose.

Don Francisco and the church official aim for the exiting door. As an educated gesture, Don Francisco opens the door for him. All of a sudden, the chief-nun appears and Sor Juana is standing right behind her, he can barely see her. The chief-nun and the church official want to direct the operation, but Don Francisco steps in:

—Your Highness, as you just said, all of these objects of art, are now of my property. It is also written in the contract that I will supervise Sor Juana's autographic signature; it is just me and the artist. Please, no one else in the room.

The church official and the chief-nun are taken by surprise by Don Francisco's assertive action; they look at each other, but the contract had to be respected. The official tells the chief-nun to leave them alone. The church official exits and Sor Juana enters the room. As the door is being closed, the menacing eyes of the chief-nun are left outside.

The sound of the closing door leaves way to an immense silence. Sor Juana finds herself in front of Francisco. They don't speak, just look into each other's eyes. Long minutes pass in absolute silence, only their breathing can be perceived. Francisco reaches for the pocket in his vest and hands her a letter he wrote describing these very moments...and the ones ahead. She takes it out of its envelope and begins reading it in silence. Francisco's voice-over reveals its contents.

*Dearest Juana Inés*

*This extraordinary moment that embraces us, is filled with words that only silence can convey. A silence so familiar to you and me; it is the essential element of the ink that writes on the white page universe, where the sensed world and imagined reality dance together as a virtuoso couple to the melody, harmony and rhythm of language.*

*As you read through these lines, your eyes have become the mirror of my soul, where my self acquires its true meaning; the highlight of my words being that there is poetry left waiting for these poets, a safe shore where they can spend a lifetime together.*

*Your letter showed despair and anguish. Lose your fear Juana Inés for you are not alone. I am here, wielding my sword to face your enemies and to break the chains of prejudice that want to enslave your mind. All of your belongings I have bought... for you; you can claim them anytime. There is no Final Period in your story, or should I say in the story that you and me are writing right now.*

*Here I am, standing in front of you, as a humble admirer of the accomplished scholar and of the beautiful woman you are. You have to live as free as a butterfly. Take my hand and come with me to*

*Colombia, under your own terms; heaven will celebrate your pen gliding once again through the white page universe, but most of all I want you by my side for the rest of my life.*
*Francisco*

After reading it, she raises her head and makes eye contact with him again. She is speechless. Francisco invites her to sit, he reaches for his vest's pocket and takes out a small jewel-box. He opens the box and brings out a gold-ring with engraved words on it. He extends his hand reaching out for her. She remains still.

Francisco then pronounces his first words,
—Please read the engravings, Francisco insists.

Leaving the letter over her lap, Sor Juana takes the ring and holds it with those very fingers that had written extraordinary lines of science and poetry. She reads the engraving inside…

*First / Dream*

and the outside …        *Nise*

She closes her hand keeping the ring; then makes eye-contact with him again.
Francisco had waited for years to tell Nise the following words,

—It is now or never Juana Inés…here, in this bag, you will find a dress. I hope it is of the right size. I have three big carriages outside the convent waiting for your belongings; it will take about four hours to load them all. Come with me to Colombia, under your own terms, you will never regret it. I will be waiting for you here in the corridor, if you don't appear once they finish emptying the room, I will understand. Nevertheless all of your belongings…are still yours and will be forever. You have my address and can claim them anytime you want. I admire and respect you, but above all, I love you my Nise and certainly will miss you if you don't come along with me.

Sor Juana does not say a word. She just takes the bag with one hand and keeps the ring within the other, stands-up, looks at him once more and goes back silently to her room. As the office door is now open, the chief-nun who has tried to listen to their conversation, enters abruptly. She asks with suspicion,
—Has she been signing all of the items Don Francisco?
Francisco ignores her and doesn't answer. He stands up to direct the workers who now come inside and begin carrying books to the carriages.

It took practically the whole morning for the workers to empty the room. The carriages are ready to go. Sor Juana is not in sight. He waits for an extra hour barely moving from a single place; she doesn't show-up. He closes his eyes and turns around heading for the exit door. He stops for a moment, looks at the convent's wall in front of him, slowly sliding the palm of his hand over it, as if wanting to retain the guarded echoes of her voice, of her footsteps.

A last look, a last image to behold, he slightly turns his head. There she is, standing in the corridor in her new dress. Francisco's breath is gone. She has her head uncovered for the first time in public; she is wearing a short haircut, her big brown eyes are enormous and so inviting, she looks radiant.
He slowly walks down the corridor to meet her, just his footsteps are heard. They now stand face to face. He reaches out for both of her hands. He notices that she is wearing his ring. Both stare at each other's eyes. Francisco's lips say the following words slowly, with an accompanying sigh and a deep sense of conviction:

—Oh my God... Oh my God...you look so beautiful Juana Inés...so beautiful...an amazing reality is waiting for us…

Juana Inés answers with the same emotional intensity:
—This is a Dream Francisco...this is a Dream...let's keep dreaming...forever...

****

These last words pronounced by Sor Juana, ended SAM's presentation. The screen stopped reflecting  his emitted Voice's wave. Lights went on in the auditorium; there was a complete silence. Never before had the Observers been exposed to such a revolutionary way of envisioning the stories' universe. Reality had been redefined.
SAM looked at the audience straight ahead waiting for the first comment to arrive. Not before long, someone within the audience stood up and said:

—SAM, you know how much your peers appreciate you; but aren't we determined by the Existence Dictum of the stories' universe to be just that, Observers? Why should we attempt to intervene in the character's life or fate? We all have seen what human nature is all about. Surely there is a lot of suffering going on, but there are happy endings too. Look at nature itself. When viewing a beautiful landscape wherever you name it, the rules of survival of the fittest, of predators and predated, prevail. The weakest chicken in the nest will be thrown out by his own brothers. Who are we, to change their fate? Aren't you going a bit too far on your scientific and philosophical dissertations of the stories' universe? In addition, we would have to give a piece of our own existence in the process.

—Further, stories that happened long ago have their Final Period that will preserve them for eternity as they began, unfolded and ended. The formation of a *parallel solar-system* with Black Celestial Bodies rewrites the whole story; in more than a way, you are coming-up with a completely new set of rules that will govern the stories' universe, bypassing the Final Period altogether and therefore the Existence Dictum. Have you thought about the reaction of the mighty Gods of Fate? Do you think they will just sit and applaud when they see Black Celestial Bodies dotting the stories' universe cosmos, replacing the Final Period?

I sincerely feel we should leave our Observer's Voice in its original state. I praise your effort in trying to help the story's fate, but I do think that your proposal goes a bit too far.

Another opposing Observer,

—SAM, I do agree with our fellow Observer. You have previously stated that being represented by a far-away echo in the character's minds is already highly comforting for that person. Why should we want to change the scenario that has operated since the origins of the stories' universe? Let the unfolding stories live through their fate and leave the finished stories as they are. My friend, I have always followed your footsteps, but what you are doing now is calling for an Observer's rebellion of the stories' universe, with unprecedented consequences.

A Supporting Observer,

—SAM, our northern star, I personally congratulate you about your commitment to further develop the Observer's role which now becomes a sort of a catalyst within the brain-mind of story characters helping crystallize those *Moments of Clarity* that everyone deserves to have. Are we justifying our confined Observers' enclosure just because the Existence Dictum of the stories' universe dictates so, or because it means a comfort zone to us? We have always submitted to the laws of the mighty Gods of Fate, who is to say that a Final Period should be the fate of every story?

SAM has given us, Observers, the possibility of Black Celestial Bodies. A story to be continued, a story to be rewritten, a story that can resolve grieve or resentment, a story embracing freedom and happiness…

A senior, highly respected Observer stands up and cries out loud:
—Good Will! Let our Voices be Voices of Good Will that should spread out to the whole universe! The eternal confinement of the Observers within the brain-mind of a story's character is over. It is our conviction, more than a commitment, to sacrifice our existence for the well-being of others. And far from disappearing we will be the air sustaining the wings of those birds freed from their cages.
He raised his arms in the air and accompanying his hands expressions, said in a clear and loud voice:
—Let's follow SAM, our northern star. Let's break the chains of destiny! Long live Black Celestial Bodies!"

The audience was now in mayhem. Oil and water separated quickly.
Those in favour and those opposing Black Celestial Bodies.
SAM tried not to lose control of the situation.

—Order!, Order! Fellow Observers! Fellow Observers!
It is not my intention to bring the Observers into a civil war!
For every conscious mind, one of the hardest things to do is to keep faithful to oneself. There are many justifying arguments that prevent us from being what we truly were meant to be.

— I fully acknowledge and respect the structure of nature. Man himself is the greatest predator on the planet of the living. I love every single expression of life, but we are the Observers of the stories' universe, where the sensed world and imagined reality ink-skate together like figure skaters on the white page, dancing to the melody, harmony and rhythm of language.
In the physical universe, the Black Hole formation is unavoidable, but in the stories' universe an inspired being will be able to have *Moments of Clarity* that will lead to the formation of Black Celestial Bodies in their own story, real or imagined, avoiding the Final Period; a black titan that swallows the whole story, its space and time, confining them to a tiny black dot held together by the Forces of Fate.

It is in this universe where the feeling of helping others has just grown inside of me. Standing tall to lend a hand to the weak, to the innocent's suffering, to the good person just about to make a fatal decision, or even giving a *moment of clarity* to a character that will allow him or her to *rewrite their own story altogether.* I will follow my soul's advice. I am aware of the cost of doing this. Every time I will pay with a piece of my own existence.

The majority of the Observers stayed in the auditorium to learn how to achieve a kaleidoscope state of mind, however many dissidents began leaving. SAM said loud and clear:

—I am proud to be an Observer of the stories' universe and to have you all as my peers; even those who do not agree with me. Self-sacrifice for others is not meant to be the *leitmotiv* for everyone.

An opposing Observer as he was exiting,
—Fools! The mighty Gods of Fate will not allow for a Black Celestial Bodies' rebellion; they will crush you!

SAM thanked the loyal Observers for their support and  next, instructed them on the meditating technique conducive to achieve a kaleidoscope state of mind.

***

## THE MIGHTY GODS OF FATE
ROYAL COURT

The chairman's fist smashes over the conference table.
—Our gravitational wave detectors have registered unprecedented low Final Period formations. What does this supposed to mean? A Black Celestial Bodies' rebellion? They are all over the cosmos of the stories' universe and are spreading fast. I told you that SAM's existence had to come to an end; it was obvious that his laboratory findings would be turned against us, but we have to learn the hard way! They want to take over. Fools!

—The Existence Dictum will prevail; how can a stories' universe ever be conceived without a Final Period? Just as in the physical universe an aging dying star will collapse forming a black hole, the Final Period is essential to the stories' universe. Stories begin and end, and they follow their own fate. These are the laws of the cosmos, nothing should be done to change them.

—The rebels are coming-up with a new set of rules, that have as their main objective to undermine our power. What do they want, happy-ending stories? A whole pink-colored cosmos?
Is this what mankind is all about? NO! *'Homo homini lupus'* *'Man is wolf to man'*. How do you visualize the history of the Roman Empire, as a series of smiles and handshakes? Go and ask Brutus while he is hiding the dagger that will bring down his own father.

—World War II had 60 million casualties, and SAM has brought the stories' universe pantheon into civil war to save the little kitten in the tree!
He raises his arms in the air and says loudly and defiantly:

—SAM, do you hear me! What about the millions of slaves that never saw their hands free of chains? Each and every one of those stories deserves your compassion and help, don't they?

—Let the human jungle live through another day; only the victorious will celebrate a new dawn. This is the way it goes, and you cannot change it!

—SAM, you are an idealist with no feet on the ground! Why are you doing this to all of the mythical Gods of the stories' universe, when you are a part of them! You are supposed to be just an Observer and report to your Superiors, is that so difficult to do? But now you have made your move, and there is a law in physics that I would like to remind the scientist of: for each action there is a reaction of the same magnitude but opposing direction. You can expect our reaction SAM, all of the rebels will pay the price with their own existence. Then we will override the Black Celestial Bodies' stories. We will find the way to do it. Kaleidoscopes are not to be found just on your side.

Chairman's assistant,
—Your Lordship, we have contacted opposing Observers that can lead us to SAM's laboratory.

Chairman,
—We need his exact location, his coordinates. We will make two neighboring Final Periods fuse in its vicinity; the immense gravitational wave thereafter will devour SAM and his whole laboratory completely. What an irony, the Final Period will mean the end for someone that is helping others avoid it. Once we get him out of the scene, we will start fixing up the whole cosmic mess he has left behind.
The Chairman raises his arms in the air again and says loudly and defiantly,

—There is no story that will avoid its Final Period. Every single Black Celestial Bodies' solar system will be wiped out!

Fate and Existence are fused together, just like hydrogen and oxygen make up water. It is this water that circulates through the veins of every being in this universe. They can not be separated and never will!

 SAM, if you think you are the only one using science for his own purposes, wait and see how *this story* unfolds!

## THE REBEL OBSERVERS' MEETING

SAM anticipated the Royal Court's mighty reaction and called for a secret meeting. There they learned that some of their fellow Observers couldn't survive the conversion of needed own mass to sustain the Voice's wave; they had given their existence for building a better cosmos, a cosmos full of hope, breaking the chains of destiny, of the Final Period and the Forces of Fate, allowing for the formation of Black Celestial Bodies.

SAM is talking to his followers:
—I feel so proud and fulfilled to belong to this courageous group of Observers. You might already know that we have lost some of our dearest and respected friends, they sacrificed their existence for the well-being of others. This is what redefines the stories' universe, *caring for the stories of others*. The cosmos is being dotted with Black Celestial Bodies avoiding the Final Period and this is just the beginning. There is no turning back, hope in this universe is here to stay.

—The Mighty Gods of Fate will strike back. By fusing existing Final Periods they will send us gravitational tsunamis to destroy us. We must disperse and act individually remaining in contact through our Voices. A universe without light or Voice, is a universe no more. Our Voices will always be there, reaching out to every corner, every shore, every story's character, every soul in this universe.

# The Story of Stories

—The Gods of Fate still have a puzzle to solve: the kaleidoscope state of mind. The kaleidoscope state of mind, so important to us now, is by its own nature elusive and best describes consciousness and its manifested creativity and imagination. How many images do you see through the kaleidoscope? One, an ever-changing image, renewing itself by the second. This is the very reason that this state of mind could break the quantum-lock and allow the Good-Will Observers to help bring a *Moment of Clarity* to the stories. In order for this to happen though, we also have to convey the message to those characters the important meaning of *your self*. From time to time our *true self* gets fragmented due to circumstances and we have to recover it, recognizing it in the mirror once again.

—There is no such thing as a Black Celestial Bodies' rebellion, as the mighty Gods of Fate want to describe.
The stories' universe owes its existence to imagination. It is your imagination that redefines reality and makes it as valid as the physical world. When this physical world, represented as a reality in our brain-minds, is placed under the microscope of neuroscience, it becomes evident that *reality is just a human illusion.*
Such a conclusion leaves us as perplexed as learning that the universe of the very small, like the nucleus and electrons of the atoms, which are the essential building blocks of matter, describe a reality totally different from our notions of the physical world we perceive through our senses, including the concepts of space and time. It is here where imagination plays a major role in our lives and defines who we really are.

\*\*\*

SAM and his fellow Observers contemplate the beauty and complexity of the Narrating Voice through the metaphor of the figure-skaters; the sensed world and imagined reality ink-skate together  on the white page, dancing to the melody, harmony and rhythm of language.

61

Neuroscience will help SAM continue to unravel the brain's intricate mysteries, but inevitably, metaphors, philosophical thinking, mathematics and wave-functions will have to come into play to reveal some of the myriad facets of human consciousness.

Finally, the Narrating Voice leaves its originating brain made up of living cells, as a light that will be the creator of yet another universe, the ethereal universe of thoughts.

In the stories' universe cosmos, the battle between the Forces of Fate and those of Free-Will has just begun. The stories' rebellion has spread fast and it has become increasingly complex; unexpected twists are bound to occur. Will SAM prevail?

***

## EPILOGUE

Dear reader, as you go through these lines, your own Narrating Voice inhabits them already; every character that plays here a role, now perceives you as a far-away echo in the back of his or her mind, SAM at the forefront. Waves generated in different locations of the same ocean, interact with each other and bring that ocean to life giving it a sense of wholeness; you are already a part of us, an Observer of the stories' universe.

The following three Black Celestial Bodies or *suspension points* next to the last word of this page, not only make it an open-ended literary work, but confluent with the rest of the stories in the stories' universe; in this context, *your own story has just added one more...*

\*\*\*

**\* (1) Additional information on Sor Juana's referenced text.**

\* *Full names of historic characters*:

-Francisco Álvarez de Velasco y Zorrilla, *Colombian poet* (1647-1708).

-Manuel Fernández de Santa Cruz, *Bishop*

-Antonio Núñez de Miranda, *Sor Juana's Confessor.*

(Cruel irony, Miranda also served as an *advisor for the Inquisition*)

-Francisco de Aguiar y Seijas, *feared Bishop*

-Antonio Vieyra, *Portuguese Theologist; Aguiar's doctrinal stronghold.*

(1) Historic landmark reference:

"El Poeta Colombiano enamorado de Sor Juana"

"The Colombian Poet that was in love with Sor Juana"

by José Pascual Buxó

Link: http://www.bdigital.unal.edu.co/1317/5/01PREL01.pdf

Important Note:

*Francisco's letters to Sor Juana, Sor Juana's answering letter* as well as the conception and dialogues of their encounter described in this Black Celestial Bodies' story, *are fictitious.* They were written by the present author, Sergio Sánchez Cordero.

For further Reading:

-"Sor Juana Inés de la Cruz o Las Trampas de la Fe"

 "Sor Juana Or The Traps of Faith" (English translation available).

By the 1990 Nobel Laureate Octavio Paz. Sor Juana was a key figure in his literary work.

Additional notes:

-Sor Juana (1651-1695) and Francisco Álvarez (1647-1708), are poets that were born and died around the same dates.

-Her flagship poem *'First I Dream'* (Primero sueño) is also known as *'The Dream'* (El Sueño).

-The word Cruz in English means cross, and it was often used as a part of a name in members of the catholic church.

-The prefix Sor is used by the catholic church's nuns.

## About the Self-Publishing Author

New York University Master of Science, class of 1978. New York University College of Dentistry, PostGraduate in Periodontology 1975-1977. Basic science publications in microbiology and clinical periodontology. Holds patents in the USA and Mexico.

### Other books by the same author available in Amazon.com
*(Also printed by CreateSpace, an Amazon.com company)*

### 'The Day of the Children's Crowns'
*The story that will change a centuries'old tradition (2017).*
This book proposes an important addition to the milk-teeth shedding tradition during elementary school, establishing *a new celebration* where the Tooth Fairy and her assistant Teethy Mouse become collaborative heroes in preventive dental care; *the centerpiece of the story* being the *crucial 4 first permanent back-teeth*, that make their debut at 6 years of age and are not announced by a shedding milk tooth in the process (this is the main reason they go unnoticed). Contains a CD with orchestated music and a children´s choir; 5 songs with lyrics in English, same 5 songs with lyrics in Spanish. e.g. The Tooth Fairy Song. Music and lyrics by present author.
**www.fairyandmouse.com**

### 'El Día de las Coronas de los Niños'
*La historia que cambiará una tradición de siglos (2017).*
Este libro propone una importante adición a la tradición de la caída de los dientes de leche durante la escuela primaria; establece *una nueva celebración* donde el Ratón de los Dientes Teethy Mouse y el Hada de los Dientes Tooth Fairy se convierten en héroes colaboradores de la prevención dental. El centro de la historia son los *cruciales 4 primeros molares permanentes* los cuales no tiran a ningún diente de leche al salir, y por ello pasan desapercibidos. Audio CD con música orquestada y coro de niños; 5 canciones en inglés y las mismas 5 en español, e.g. La Canción del Ratón de los Dientes. Música y letra por el mismo autor.
**www.hadayraton.com**

# OBSERVER'S NOTES

## OBSERVER'S NOTES

# OBSERVER'S NOTES

'Black Celestial Bodies'
(2016)

Printed by CreateSpace,
An Amazon.com Company


Available from Amazon.com and other book stores.